German Railway Gun

Leopold

28 cm K5 (E)

Jan Coen Wijnstok

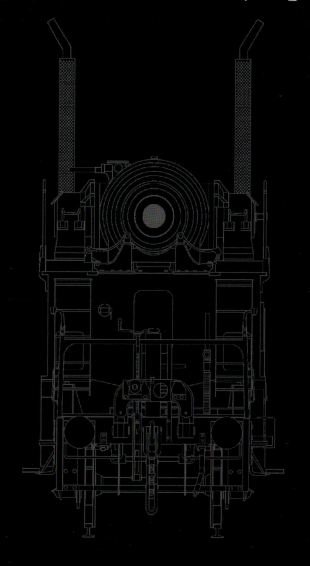

Model Centrum PROGRES

Armor PhotoGallery #12R

German Railway Gun
28 cm K5(E) Leopold

Jan Coen Wijnstok

Published by Model Centrum Progres, Poland
Warsaw, January 2015

Text © 2005 Jan Coen Wijnstok
Scale plans © 2005 Jan Coen Wijnstok
Current color photos © 2005 Joseph P. Koss, Jr. (for Aberdeen exhibit); Peter Crispyn and Henri Klom (for Audinghen exhibit)
Front cover art © 2005 Arkadiusz Wróbel
Historical photo credits: Daniele Guglielmi Collection; Jeffrey Plowman Collection; Marcel Verhaaf Collection; Imperial War Museum

English proof-reading by Joseph P. Koss, Jr.

Acknowledgments
The Author and the Editor wish to express their thanks to Dr William F. Atwater, Director, US Army Ordnance Museum, Aberdeen Proving Grounds, Aberdeen, Maryland, USA; Peter Crispyn, Deinze, Belgium; Daniele Guglielmi, Calenzano, Italy; Jeffrey Plowman, Christchurch, New Zealand; Marcel Verhaaf, Hilversum, The Netherlands

ISBN 978-83-60672-24-2

Edited by Wojciech J. Gawrych
Cover layout, design and layout by
PROGRES Publishing House, Warsaw
DTP and prepress by AIRES-GRAF, Warsaw
Printed and bound in European Union by
Drukarnia O.W. READ ME, Olechowska 83,
92-403 Łódź, Poland

Second Edition

Exclusive Distributor
Model Centrum
Warsaw, Poland
fax +48 22 6288741
wjg-books@wp.pl
www.modelbooks.republika.pl

28 cm K5 (E) Ausf. C

US Army Ordnance Museum – Aberdeen, MD.

1 2 In early 1944, the Allies were up against the Gustav line running across Italy, south of Rome. The actions around Monte Cassino are a well-known part of that campaign. To force the issue, the Allies landed a force at Anzio, north of the Gustav line. However, the invasion force was contained in its beachhead for months. Among the German forces containing it was Eisenbahnbatterie 712 (railway battery). It had been meant for the North African campaign but the Afrikakorps was defeated before its arrival. Subsequently stationed near Milan in the north of Italy, they were nicely at hand. The two guns of the battery, named 'Leopold' and 'Robert' by their crews, bombarded the beachhead from the safety of railway tunnels. They became known collectively as 'Anzio Annie' by the Allies.

When the German army finally had to retreat, the guns had to be abandoned because bombing destroyed the railway. The crews tried to destroy their guns, but only damaged 'Leopold'. Eisenbahn-batterie 712's personnel escaped and they were reequipped with new K5's. The American 168th Infantry Regiment captured the guns. 'Leopold' was subsequently shipped to Aberdeen Proving Grounds, one of the US Army's test facilities, for assessment.

The official designation of 'Leopold' is 28 cm Kanone 5 (Eisenbahn), abbrevia-ted to 28 cm K5 (E). This literally trans-lates into English as 28 cm Cannon 5 (Railway). This particular gun is an Ausführung C (Model C), with the aiming platform in a low position. Its transport cover doubles as a roof when in use.

K5's were built by Krupp and Hanomag.

Without a manufacturer's plate and date, it is impossible to know when and where the gun was built. Since it has a relatively low carriage number, it probably is an early gun. 'Leopold' is reasonably complete. The generator unit, which is missing, would not have been on the gun since it was in transport mode when captured. It was probably never shipped. The woodwork has completely rotted away on the work surfaces of both the gondola and the front railway truck. 'Leopold' was overall Dunkelgelb (dark yellow), with white lettering, when it was captured. It has been repainted since but the lettering is accurate, if incomplete. There should be a lot more, both on the gondola and on the railway trucks. The latest coat of paint is camouflage, which is not historically accurate. Period photographs show camouflage made up

of broad bands of color, probably the paint scheme used between 1935 and 1939, which was made up of 2/3 Dark Gray and 1/3 Dark Brown (see photo on page 41).

3 Overall view of the front railway truck. It is a twelve-wheel truck, consisting of two pairs of three bogies. The arched coupling between the buffers is used in conjunction with the Vögele Drehscheibe (turntable). It consists of the Schiessbettung (firing bed) and a circular set of rails, which gives the gun a 360-degree arc of fire. The coupling secures the gun to the firing bed. The turntable was transported on two special railcars, one for the firing bed and one for the rails. One was actually captured with 'Leopold', but presumably not shipped to the United States. Note the open hatch, it gives access to the box that makes up the side of the gondola.

4 The hinges of the firing bed coupling are riveted to the top and bottom of the railway trucks bed (see photo [175]).

5 The top of the coupling; note the hook used to secure it. Next to it is the threaded part of the handbrake operating rod.

6 Overall view of the front of the railway truck. The firing bed coupling is in the middle. The vehicle coupling and the operating mechanism for the handbrake are situated within its arch. To the left and right of its hinges are angle cocks for pressure hoses. There were two separate systems, one for compressed air running through the left side of the vehicle and one for steam on the other side. The compressed air system was in use on 'Leopold' as can be seen by the surviving hoses. The steam system would have been used in conjunction with a steam locomotive.

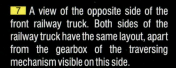

7 A view of the opposite side of the front railway truck. Both sides of the railway truck have the same layout, apart from the gearbox of the traversing mechanism visible on this side.

8 The winch to lower and raise the firing bed coupling. The crank once had a wooden handle. Note the ratchet on top of the crank.

9 Overall view of the end of the railway truck. The handle of the handbrake is situated to the right of the winch. Wooden slats used as a work surface would have concealed the openings in the deck.

10 Right side of the winch with cable drum. Part of the gears can be seen between the plates. The slit between the winch and the back of the handbrake serves as a guide for the cable.

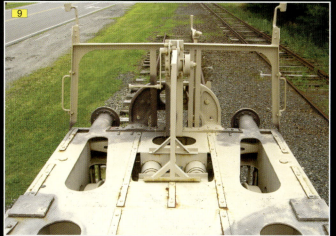

11 **12** **13** Three views from different angles of the brackets used to hold a lantern or possibly a square color-coded plate, to denote the last carriage of a train. The plates measured 25 x 25 cm and the back was completely white. The lanterns were about 50 cm tall oil lamps. They came in different shapes, though always rectangular. The lantern had the same color-coding as the plate on one side. The door was filled with red sheet glass and the back with clear sheet glass.

© Jan Coen Wijnstok

14 Side view of the railway truck. The plaque next to the winch is not original, but a museum addition.

15 16 Views through the holes in the railway truck's bed. A wheel and parts of the brake system can be seen.

17 18 Parts of the pressure hoses that run from the front of the gondola to the railway truck (see photo [33]).

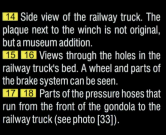

19 Another view of the deck of the railway truck. The strips running lengthwise were used to attach the wooden slats of the work surface.

20 View of the railway truck from underneath the gondola. Its bottom can just be seen in the top of the picture. To the left and right are the hangers for the brake shoes. They are joined at the bottom to a rod that connects to the central brake mechanism seen in the middle. On the top left is a hook for attachment of the chain connecting the railway truck and the gondola in transport. This chain would have prevented the railway truck from shearing when derailing. Keeping the trucks in a straight line would minimize damage in case of accidents. Bear in mind that K5's were able to use bad or quickly laid temporary track, so accidents were taken into account.

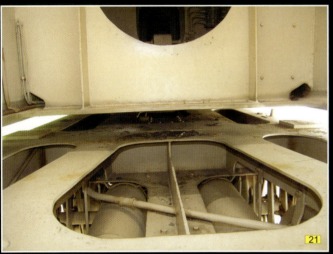

21 22 23 Several views of the pressure vessels for the brakes. The pressure drives central brake cylinders that operate the brakes through a system of rods and hangers.

24 View of the rear of the railway truck with the pivot and traversing mechanism. The main aiming of the gun was done by means of either a curve in the rails or the turntable. For fine adjustment, the gondola could be traversed about two degrees to either side.

25 26 The back of the traversing mechanism is seen here. The box in the middle is a switchbox that turns off electrical power when maximum traverse is reached. The arm, of which there should be one on either side, is the actual switch.

27 Close-up of the slide and rails of the traversing mechanism.

28 This structure is probably storage for cleaning rods and the like.

29 One of the wheels that keep the gondola level when turning.

30 This mechanism is probably a locking system for the traversing mechanism. The flaps on the end can be turned inwards and secured by wing nuts thus preventing movement. This would normally be done from the operating panel on the side of the railway truck with a bolt in the slide that also interrupted electricity. This may be a field repair since it can be seen in wartime photographs of 'Leopold', but not in any other example of K5.

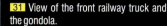

31 View of the front railway truck and the gondola.

32 The front of the gondola. The hatches behind the ladders give access to the hollow sides of the gondola. The ladders themselves lead to the top of the gondola and the barrel clamp. Next to them are couplings and hoses for compressed air and steam. The middle hatch opens into the front compartment that houses the motors, gearing and such for the elevation of the gun.

33 One of the hoses for compressed air. This one was probably cut when the crew tried to destroy the gun.

34 35 36 Different views of the barrel clamp. One part has broken off and the other one was bent straight. When not in use, the movable parts of the clamp can be secured to attachments on the extreme ends of the clamp's base.

37 Overall view of the gondola. A round bolted cover and levers can be seen on the front left side. The round cover closes off the access hatch for installing an auxiliary motor in case the generator does not function. The levers are for hand operation of the elevation brakes, which are normally operated electrically. This is an automated system to prevent the barrel from dropping when the equilibrator fails.

38 39 Close-ups of the gondola. The name was originally painted on both sides of the gondola

40 41 The chain used to connect the end of the railway truck to the gondola in transport mode. There was one on each end of the gondola. Note the neat stowage when not in use (see photo [20]).

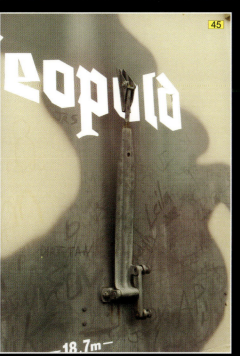

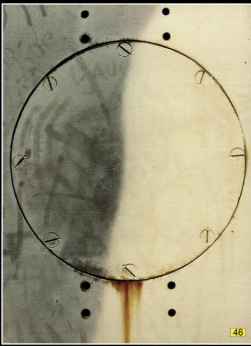

42 The left side of the gondola.

43 This photo shows the cover of the aiming stand, it folds up and doubles as a roof when in use. Note the handles on the bottom edge. A platform to stand on can be attached to pegs beneath the cover (see photo on page 45, top).

44 One of the ladders that give access to the top of the gondola. The other is directly opposite. The bottom part can be let down to reach the track.

45 This is the indicator to measure how far the gun has slid back after firing. It was folded down and secured. A ruler was fixed to the track beneath it. Before the next shot, the gun could be brought back in its initial position, or the crew could compensate for the amount of travel in aiming the next shot.

46 Cover of a manhole that gives access to the connecting axle of the equilibrator and the elevation operating rods, allowing their bolts to be serviced.

47 48 Two views of the breech and its surroundings. Behind the breech is the Ladeklappe (loading flap). It would be down when loading the gun and accommodated the loading trolley with the grenade and the propellant charges. It is then raised to give room for elevation and recoil. The entire platform was originally covered with wooden slats to provide a work surface. Note the size of the projectile; rotating ribs can be seen on its side.

49 The underside of the loading flap. It is hinged at the back and a counterweight is attached to the center of the hinge rod to keep it in the up position.

50 51 More views of the loading platform. Note the foot and upright of the railing that could be folded down. The railing would have run all along both sides of the gondola. Large parts are missing.

52 A good view of the cartridge catcher. The stand for the rammer should be behind it. This rammer was used to release the breechblock when it became jammed. The container in front is for stowing it (see photo [171]).

53 **54** **55** Three views of the car-tridge catcher. The tray was swiveled behind the breech after firing to hold the metal cartridge from the main propellant charge. Note the catch that holds the loading flap in a level position. The metal strips with holes are for attachment of wooden slats for the work surface.

56 57 Looking to the rear at the breech, where it comes up against the gun cradle. The welded strips are not original and probably keep the barrel from sliding back.

58 A nice view of the breech and the breechblock operating mechanism. Note the catch for the loading flap, with retaining pin, in the bottom of the photo.

59 The breech from below.

60 Close-up of the breech with manufacturer's codes for different parts. The date '1941' only means that the breech was manufactured then. Parts of the gun could have been made by different manufacturers or later exchanged. The code 'bye' is the one for Hanomag.

61 The gun barrel. This is either a T10 or T7 barrel, which has 12 rifles with a depth of 10 and 7 mm, respectively. They used grenades with rotating ribs instead of the conventional rotating bands. The rifling depth of 10 mm was reduced to 7 mm because the barrels were prone to bursting. T10 barrels were reinforced and used up to the end of the war (see page 43).

62 63 64 65 66 67 Six different views of the breechblock operating mechanism. It moves the breechblock sideways to open and close the breech. The gears can be clearly seen inside. The cover is missing, as is that on the round hole facing front (see photo [67]). The crank once had a wooden handle.

68 69 The sheet metal covers on both sides of the gun-cradle.

70 Another view of the aiming stand. Instruments and the mount for the gun sight are behind the cover. When in operation the gun sight was connected to the trunnion with a rod to indicate elevation. This rod passes through a hole in the top left of the cover. Elevation and traverse were also controlled from this position by hand wheels. The trunnion in its mount is seen above, on top of the gondola. An angle cock is situated beneath the aiming stand; it was used in transport mode to connect the pressure hose to. The hose hanging to the right was used in firing mode.

71 The tread plate on top of the trunnion retaining plate made lubricating the trunnion easier.

72 A close-up of the trunnion; the trunnion cover is missing.

73 The bolts on the trunnion retaining plate. The brackets seen here could be used as steps.

74 75 These photos show the way the trunnion connects to the gun-cradle. Note the rain covers on the inside and outside of the trunnion retaining plate.

76 A good view of the middle section of the gondola, with the indicator for recoil slide in the middle and the manhole cover to the left. The gun-cradle is seen above, including a good view of the side cover.

77 The gun-cradle is a one piece casting. Its front end is seen here.

78 79 80 Three views of the electrical system and its connector box on top of the gun-cradle.

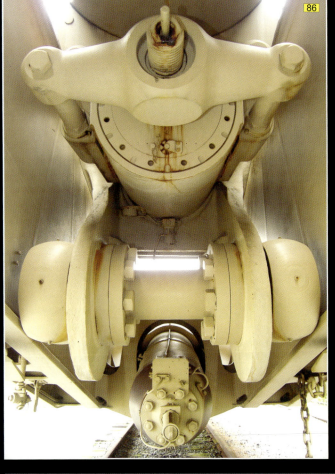

81 82 83 84 85 The lower part of the gun-cradle seen from different angles. The smaller cylinders on the outside are the recoil brakes; the large one in the middle is the recuperator. The latter brings the barrel back into position after firing.

86 87 Two views of the rear end of the recuperator. The rods alongside it connect to horns on the breech. Both brakes and recuperator work together to slow recoil and reposition the barrel. Beneath it is the connection between the cradle and the elevating mechanism. The ends of rods running forward are seen head on. Behind and beneath this is the end of the equilibrator.

88 A good view of the horns on the bottom of the breech. Both the recoil brakes and the recuperator are attached to them. The bolts at the top are from the recoil brakes.

89 90 Looking into the gondola, the connecting rods and camshafts of the elevating mechanism can be seen, as well as the front of the equilibrator. Note the way these are connected. A slide, that keeps the whole assembly from going all over the place when elevating, runs along the rail in the middle. The part in the middle of the slide should be bolted to the front of the equilibrator, which it is not on 'Leopold'. Connecting rods to the cradle run to the back and camshafts to the front.

91 The other end of the rail is bolted to the inner bulkhead of the front compartment that houses the rest of the elevating mechanism. The underside of a camshaft can be seen.

92 93 The rail from below, the sides of the slide can be seen gripping the rail. This rail is fixed to the insides of the gondola with cross members.

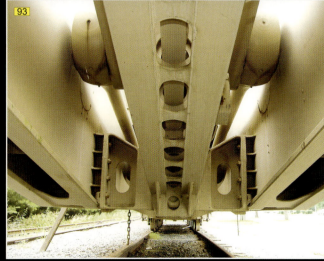

94 Another view of the front end of the equilibrator. Its attachment to the insides of the gondola is visible beneath the cross member. The equilibrator should be connected to the slide, but is not. The bolt holes for it can be clearly seen. This is the head of the inner cylinder of the equilibrator, which comes out when the barrel is elevated.

95 96 97 The rear end of the equilibrator is seen sticking through another cross member with lightening holes. The equilibrator keeps the gun balanced when elevating or lowering and reduces the force needed to do this.

98 99 100 Views of the underside of the gondola. Note the grease nipples on the ends of the connecting rods.

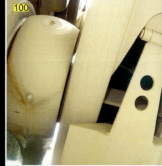

101 Access hatch into the rear compartment of the gondola from underneath the gun. The object seen in the top of the hatch is the counterweight for the loading flap, which is right above.

102 Another look at the counterweight; a hatch can be seen above it. Note all the stiffeners on the roof members.

103 Overall view of the rear compartment; note the inner bulkheads. The box in the middle is a fuse box. The small box on top of it and the shaft going into the hollow side of the gondola are probably used to manually raise the gondola. A hand wheel to operate it is on the floor to the left of the box.

104 105 Examples of rather incomplete fuse and switch boxes. They would have had lids originally.

106

107

108

109

106 A good view of the rear of the gondola.

107 108 The top of the rear compartment. The hatches would have been hidden by the wooden work surface. The rails for the loading trolley are running along the middle. Rammers for the projectile and the main propellant cartridge were stowed between them when firing. Their attachments are missing.

109 The split access hatch for the rear compartment. The space between the gondola and the generator truck is not big enough to open and close a single hatch. Electric cables were run through the hatch to power the gun.

110 111 Close-ups of the angle cocks for the pressure hoses. The levers were used to open or close the connection.

110

111

112 113 114 115 The turntable on the rear railway truck. In transport mode it would be completely under the gondola and thus not visible. In order to make room for the gun in recoil, the railway truck was pulled backwards 1.9 meters, as is the case on 'Leopold'. The gondola was raised slightly to do this and lowered again when in the aft position. This also made room on the rear railway truck for the generator unit. The steel cable attached to the step of the gondola is not an original feature.

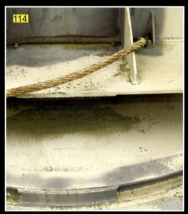

116 The lightening holes in the deck of the railway truck also provide access from the top. There were no wooden work surfaces on the rear railway truck. Note the rails for the generator unit running between the holes.

117 118 119 Different views of the wheels protruding from the bottom of the gondola. These were used to stabilize the gondola in curves. There are two sets, one in front to be used during transport and one at the back used in firing mode.

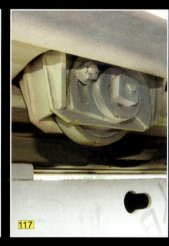

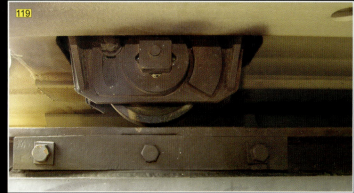

120 121 Overall view of the rear railway truck and gondola. Mechanisms used to lock the railway truck in the forward or aft positions are above the stabilizing wheels on the gondola. These would be operated with a large spanner. The arms would then be secured with the pins on chains seen in the picture. The top securing points are seen above the operating mechanisms. 'Arm down' is the locked position, so the rearmost one should be down and the foremost one up in this mode. Since none of the wheels are touching the turntable, we can deduce that the gondola was not lowered again after the railway truck was pulled back. The two sets of three bogies that make up the running gear of the railway truck can be seen here. Only the outer two sets of wheels have brakes. The museum data plaque on the end is not original; it would interfere with the generator unit.

122 The inner end of the rail carriage.
123 The outer wheel of the three-bogie set. The axles slide up and down in the frame seen behind the leaf spring. Note the journal box on the end of the axle. It bears the text: Schütte, Meyer & Co Letmath 1939.

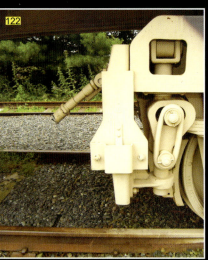

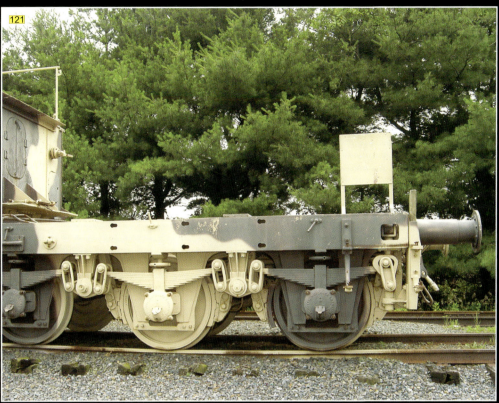

121

124 The construction that keeps the leaf springs in line. A rod running across the width of the truck connects the outer ones on both ends of the railway truck.

125 The middle wheel.

126 Close-up view of the construction that keeps the leaf springs in line.

127 The inner wheel. There are three levers in this picture. The one on the left is probably part of the hydraulic system to lift the gondola. The one on the right is used to switch brake pressure from goods to passenger train. The one above is part of the locking device for the springs, whose main part is above the bracket in the middle of the spring. These would be completely down when in firing mode. The suspension of the gun had to be locked to prevent damage from the recoil forces.

128 The box containing pressure gauges for the hydraulic pump used in raising the gondola.

129 130 131 More views of the running gear. Photo [129] shows the attachment for a hand wheel; another one is located on the opposite side.

132 The end of the railway truck with buffers and the brackets for steps. These would have been wood; the bottom one was slightly wider.

125

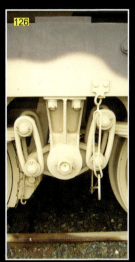

126

127

131

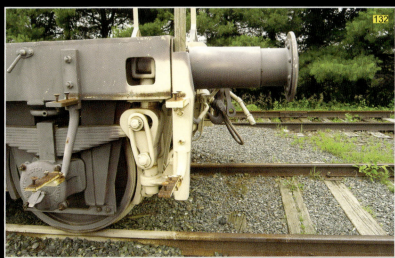

132

133 The inner end of the railway truck. The steps in the middle lead to the round access hatch into the rear compartment of the gondola. Note the coupling for compressed air; most of the hose is missing. It connected the gondola and the rear railway truck, and could be uncoupled when pulling the railway truck back. It would have been reattached once this operation was completed.

134 Upper parts of the central operating system for the brakes.

135 Details around the brake shoe hanger. Note the hook for the chain connecting railway truck and gondola when traveling. Under it is another coupling, probably for steam, since the one for compressed air is on the other side.

136 Lower parts of the central operating system for the brakes.

137 A nice view of the underside of the railway truck. The central brake system components can be seen in the middle behind the steps. Next to it on both sides are pressure vessels for compressed air to operate the brakes. Note also the attachment of the brake shoes.

138 139 Details of the brake shoe hangers.

140 141 More views of the railway truck's inner face.

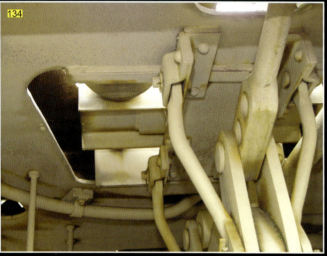

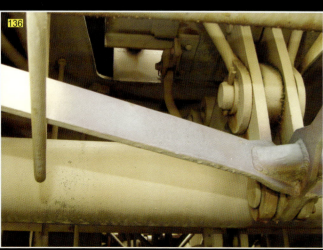

137

138

139

140

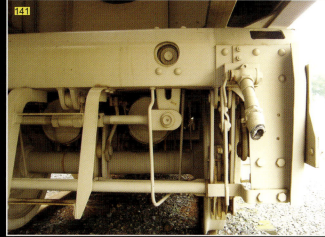

141

142 143 The right rear views of the rear railway truck and gondola. These pictures make clear that both sides of the gun are essentially the same. The gun is operated from the other side, so this side is simpler in layout. Note that the name 'Leopold' should also be on this side of the gondola. The ladder on this side is missing.

144 The rear buffers. The buffers can be shortened or lengthened by rotating them. They are sprung on the inside to cushion the impact of pulling out or braking. Take note of the hinged lower step on the railway truck.

145 146 147 148 Detail views of the sides of the railway truck and the running gear. The hinge on the lower step can be clearly seen in photo [145]. Photo [148] shows the attachment for a hand wheel or crank. Another one is located on the opposite side.

149 The gap between the inner pairs of wheels is wider, hence the different attachment for the springs. There is no connecting rod across the railway truck as in the attachment shown in photo [148]. There are no brakes on the inner wheels.

142

144

145

146

151

152

153

154

143

150

150 This frame would hold the transport authorization, stating purpose and destination on a form glued beneath the chicken wire cover.
151 152 153 154 155 156 157 158 More details of the side and running gear, finishing with the inner end of the railway truck. The wheels have a diameter of 900 mm. Normal Deutsche Reichsbahn (Imperial German Railways) wheels were 940 – 1000 mm. All the other parts of the railway truck and running gear conform to Reichsbahn standards.

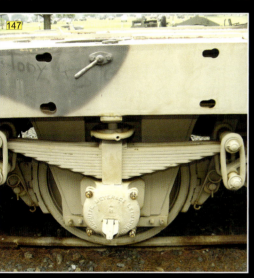

147

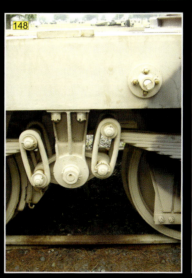

148

149

155

156

157

158

159 Overall view of the rear of the gun. The ladder extends quite far from the side of the gondola. The ladder consists of three parts that can slide up to shorten it. The outrigger can be collapsed and the whole assembly stowed on the side of the gondola with two hooks seen in the picture.

160 One of the brackets for the steps on the railway truck. The wooden steps have rotted away. The lower one would have been raised in transport.

161 Another good view of the buffers and coupling. The layout is basically the same as on the front railway truck.

162 The side of the coupling, showing the draft hook.

163 164 165 Head-on views of the front of the railway truck. The angle cocks for the air and steam hoses are next to the buffers, only the air hose is in use. Above them are the brackets and retaining pins used to secure the rails for pulling the generator unit onto the rear railway truck. The generator was transported on the Munitionszubringerwagen (ammunition supply car).

166 The front of the railway truck, seen from the top. The couplings on 'Leopold' are not complete; the screw is missing. It should be connected to the yoke, i.e. the bars with holes seen hanging from the draft hook. The staple, similar to the one hanging down would be connected to the other end of the screw. Together this forms an adjustable system for coupling railcars (see page 34 for a complete system). The hook and staple hanging from the draft hook are an additional safety measure. Both systems would be used in coupling.

163

164

165

166

167 The top of the rear railway truck. The rails for the generator run along it. The rectangular plates with keyhole slots play a role in fastening the generator to the railway truck. Triangular structures were attached here to which the generator truck was fastened (see photo [212]). The hole in the middle shows one of the brake cylinders.

168 These are the draft springs, which compensate for the inertia of the rail cars when pulling away and create smooth movement. The draft hook attaches to this assembly.

169 170 More views of the brake cylinder. One of the angle cocks for the steam hoses can be seen in the bottom right.

28 cm K5 (E) Ausf. D

Musée du Mur de l'Atlantique – Audinghen, Pas de Calais

171

171 In 1980, a 31 m long railway gun was discovered behind a factory building of the state artillery workshop in Tarbes, in the south of France. It turned out to be a 28 cm K5 (E). There was interest from quite a few museums, but in the end it went to the Musée du Mur de l'Atlantique (Atlantic Wall Museum) on the Channel coast. Back in a place where K5's were frequently used to shell Great Britain from prepared positions, complete with bunkers for protection and hiding. How or why the gun wound up in Tarbes is unknown. Not even where it came from is clear, best guess is that it is one of the guns from Eisenbahnbatterie 749 that were overrun by the Allies near Montelimar, in the Rhone-valley. The guns stationed on the Channel coast re-treated to The Netherlands in September 1944 and were destroyed there. The gun was manufactured by Krupp in 1941, when a total of seven guns was deli-vered. It is an Ausführung D (Model D), which has the aiming stand in a high position. There was no transport cover for the aiming stand, as on 'Leopold'. The gun was repainted green and all the lettering has gone so no carriage number is known.

172

172 Richtstand Neuer Art, or new model aiming stand. Note the way the instruments are arranged near the trunnion. This is a much simpler layout than on 'Leopold'.

173 Head-on view of the gun. Note that the barrel clamp is complete and resting on its stowing attachments.

174 The large coupling for attachment to the Vögele Drehscheibe (turntable).
175 The riveted plate that attaches the coupling's hinge to the upper deck of the railway truck. There is a similar attachment on the bottom (see photo [4]).
176 A good view of the back of one of the buffers. The head is bolted to the shaft. The buffer has a diameter of 370 mm, which is standard for Reichsbahn equipment.

177 178 179 More views of the buffers and couplings. Photo [177] shows the attachment of the hinge to top and bottom of the railway trucks deck. The vehicle coupling is complete on this gun. The screw is in place between the yoke and the staple. The screw was used to tighten the coupling. The safety hook and staple can be seen beneath the vehicle coupling in photo [178].
180 The arm from the handbrake; the operating shaft is missing (see page 4).

181 The electric motor for the traversing mechanism.
182 **183** Two photos showing both ends of the electric motor.

184 **185** Nice views of the overall shape and position of the electric motor for the traversing mechanism. The operating shaft goes through the side of the railway truck on the right. The connection to the guns electrical system is on the underside of the motor.
186 This is probably a fuse box for the electrical system.
187 **188** The shape of the electrical connection can be clearly seen in these photos.

189

190

191

192

193

194

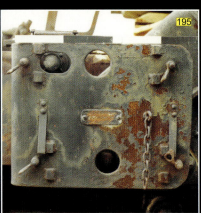

195

196

197

198 199 200 The front of the traversing mechanism. Note the sliding cover for the operating shaft.

201 202 One of the wheels that support the gondola.
203 The back of the traversing mechanism. The switchbox in the middle has retained both its arms. On the ends of the slide are metal bars that activate the switches; the one on the right is missing.
204 The side of the slide on which the gondola rests.

189 190 The side of the front railway truck. This is a twelve-wheel truck, with bogies arranged in two sets of three.
191 192 The operating panel and gearbox for the traversing mechanism. Traverse was operated electrically from the aiming stand, but it could be done manually from the operating panel.
193 194 The gearbox, seen from the front and the top.
195 The operating panel seen from the front. The traversing mechanism was locked and unlocked manually from this panel. When traversing electrically, a metal strip held in place with the fasteners in the corners, was used to block the holes. This prevented manual operation. When either a hand wheel or crank handle was inserted, power was cut to prevent damage to the electrical components.
196 197 The top and front facing side of the operating panel.

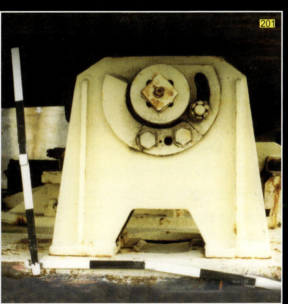

205 206 The brake cylinder from the top and the bottom. There where two in each railway truck.
207 208 This object is also part of the brake system.

209 210 211 This is probably a container for oil, since it has a draining plug on the bottom. There seems to be a lead from the top going into the railway truck.

205

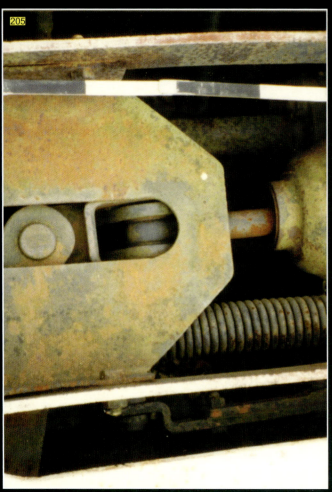

207

208

209

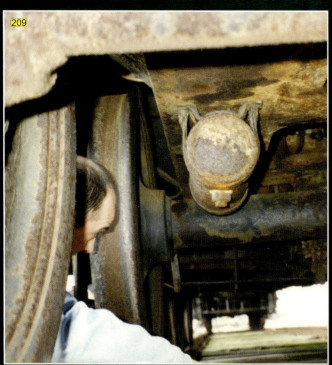

210

211

212 The steps on the rear railway truck. Note that the bracket on the left is rectangular in section, whereas it is round on 'Leopold'. One of the securing points for the generator unit is above the steps. The insert that was fastened to the railway trucks bed is missing. The generator was raised on rollers and then moved from the ammunition supply car onto the railway truck with pulleys fastened to the rear of the gondola and consequently lowered in place.

213 214 215 216 Several detail views of the running gear and suspension. The manufacturers name on the journal box reads: Knorr Bremse AG Volmarstein 1940. The axle revolves in the journal box, which is packed with oiled material. The little lid on it is for oiling the packing. Note the locking mechanism for the spring in the top of photo [215].

217 218 The top of the generator unit with the loading trolley. It has a double set of rails that join on the rear deck of the gondola. There would have been a wooden work surface on it.

219 220 The rear of the generator unit with the ammunition crane. The generator itself and its instruments are gone, though. The metal grille walk surfaces are missing from the crane. The hand wheel was used to rotate the crane, whereas hoisting was done electrically.

221 A view of the side of the generator unit. The operators were on the opposite side. The exhausts are missing on both sides, as is the generator itself.

222 The foot of the ammunition crane. **223 224** The front of the generator unit with doors for the electric leads into the gondola.

The 28 cm K5 (E) was the most widely used and best railway gun in the German Army. Development started with Krupp in 1934. The first guns were completed and test fired between 1937 and 1939. Deployment of the guns followed from 1940 onwards

Development

The principle of making heavy guns mobile by placing them on purpose-built railway carriages originated in the second half of the 19th century. Advances in metallurgy meant that ever-heavier guns could be built, but now transport became difficult, if not impossible. The railway was the only means of transport capable of dealing with such weight. By about 1900, there already was an extensive railway network in Europe, which made railway guns increasingly effective and attractive. Railway guns came into their own during the First World War, when they made it possible to give heavy artillery support over long periods of time and in widely dispersed sectors of the front. Since railway guns could be moved and redeployed relatively quickly, and were deployed tens of kilometers behind the lines, they were almost undetectable by the means of reconnaissance available at the time. Most nations used naval guns in combination with purpose-built carriages to make up their railway guns. The Germans developed a completely new gun to bombard the French capital, hence its nickname 'Paris Gun'. After the Armistice of 1918, the German Army was severely restricted in numbers and armament. Their heavy artillery and railway guns were scrapped. With Hitler's rise to power in 1933, a program of rearmament was started. High on the Army's wish list were railway guns. To provide guns quickly, Krupp married naval guns to carriages of World War One design for which they still had drawings. These ranged in caliber from 15 cm to 28 cm, but were hardly cutting-edge technology. So from 1934, Krupp started the development of completely new railway guns. One, known as 21 cm K12 (E), was derived from the 'Paris Gun'. The other, which was a new development, was designated 28 cm K5 (E). Krupp made use of extensive theoretical studies done in the 1920's and 1930's to develop a new barrel and a new projectile. In order to achieve the very long range that was requested by the Army, the new projectile had rotating ribs made of iron instead of conventional rotating bands made of copper. These rotating ribs ran lengthwise along the grenade at an angle of 5.5 de-

[Continued on page 43]

[Top] This is carriage number 919217. These numbers were carried on both sides of the gondola and on the railway trucks. It can also be seen on the firing bed beneath the front railway truck. The triangular pieces on the lower front of the cradle are the bottom attachments for the Sprengwerk; the others are on top of the gun-cradle. The gun has the two-tone camouflage used between 1935 and 1939, consisting of broad bands of dark gray and brown. There is no camouflage on the firing bed or the generator unit. The circular rail of the turntable is nicely camouflaged with branches. This photograph was probably taken during test firing or training. [Marcel Verhaaf Collection]

[*Right*] This is the first K5 built, photographed in the factory. The first pattern traversing mechanism can be recognized from the hand wheel on this side. Very long and heavy barrels tend to bend under their own weight. The construction around the barrel, used to counter-act this tendency, is called the Sprengwerk. It was used on the first three guns only, and then discontinued because it was not needed. The gun is clearly still under construction, as the railings have not been installed. [Marcel Verhaaf Collection]

[*Below*] Here is a good view of one of the early guns on the firing bed. Provisions for the attachment of the Sprengwerk are on the cradle. The Ladeklappe (loading flap) is in the upright position, so the gun must be ready for firing. The feet of the men aiming the gun are visible to the right of the rear railway truck. On the left is the special car used with the turntable to transport loading trolleys from the ammunition supply car to the gun. It seems to carry an empty cartridge back to the supply car. Another empty loading trolley is on the loading platform of the gun. [Marcel Verhaaf Collection]

[Left] Here we see carriage number 919214. There is no provision for the Sprengwerk. When or where this photograph was taken is unknown. Since it has uniform dark gray camouflage it must be after 1939. On top of the loading platform is one of the rammers for loading the gun. The light diagonal line going from the breech of the gun to the left is the lanyard for firing the gun. [Marcel Verhaaf Collection]

[Bottom] 'Leopold' seen in transport mode, the rear railway truck almost reaches the aiming stand. The cover of the aiming stand should be closed. [Marcel Verhaaf Collection]

[Continued from page 41]
grees, corresponding exactly to the rifling in the barrel. They provide the gas-tight fit that produced a higher muzzle velocity and thus a longer range. From the beginning, there were problems with the rotating ribs that regularly broke off. When the barrels also showed cracks during test firing, the guns went back to Krupp. No definite cause could be found, and in the end the chamber was strengthened and the rifling depth reduced from the original 10 mm to 7 mm. This stopped the problems with the barrels, but not with the rotating ribs, which then led to the development of a conventional projectile with rotating bands and a corresponding barrel. By 1942, there were three types of barrel in use with K5. The original one with 12 rifling grooves of 10 mm depth, called 28 cm K5 (E) T10, for Tiefzug or deep rifling. The 28 cm K5 (E) T7, with 12 rifling grooves of 7 mm depth; and the 28 cm K5 (E) V, for Vielzug or multiple rifling. The latter had 60 rifling grooves and used a projectile with rotating bands. The different types of barrel stayed in use as long as there was ammunition for them. A note of 23 May 1944 by the General of Artillery states that the two guns of Railway Battery 713 will have their barrels replaced when they have fired their remaining rounds for 10 mm rifling. The final development on K5 was the Peenemünder Pfeil Geschoss (arrow projectile). A 31 cm smoothbore barrel was developed for it. The development of the projectile itself was plagued by problems and delayed until the very end of the war. Two smoothbore K5's seem to have fired on Maastricht in the Netherlands and Verviers in Belgium.

K5 saw a few other changes as well. Very long and heavy gun barrels tend to bend under their own weight. To counteract this, a boxlike structure, called a Sprengwerk, was fitted around the barrel just forward of the cradle. There were triangular protrusions on the lower front of the cradle to attach this Sprengwerk. When this turned out to be unnecessary, it was no longer used. Provisions for it were discontinued after the third gun. The traversing mechanism was probably changed, after the first gun was built, to
[Continued on page 45]

[Right] This is a good view of the loading platform. There is a single pair of rails for the ammunition trolley. The loading ramrods can be seen in their stowed position between the rails. These and other loose pieces of equipment were stowed in a boxcar when the gun was transported. Another rod is positioned between the cartridge catcher, swiveled into position behind the breech, and the loading trolley. What exactly is taking place is not clear. One guess is that the crew is extracting a stuck cartridge. Note that many of the gun crew wear overalls. The gun is on a very narrow sunken track. This will account for the fact that the aiming stand cover is closed and the ladders are flush with the side of the gondola. [Daniele Guglielmi Collection]

[Right] This is carriage number 919214 seen from the other side. The layout of the aiming stand's instrument panel can be seen quite well. The soldier on the right is setting the sight. Beneath it is the hand wheel that controls the speed of traverse and elevation. To the right of the instrument panel is the three-way valve for the pressure system. The platform is detachable and hangs from pegs on the gondola; note the folding step. The two men on the left have folded down the indicator for recoil slide. There is a white strip on the firing bed that could be the ruler. The pressure hose attached to the rearmost cock is in the far right of the photograph. Details of the firing bed can be seen very well. [Marcel Verhaaf Collection]

[Below] This is a very good view of the generator and crane in action. The railings on the generator unit are small; the large gap was necessary to accommodate the ammunition trolley. Note that it can be closed with chains. The instrument panel of the generator can be seen clearly. The inside of the door is padded and probably white or off-white. The carriage number on the generator is 919201. The wheels of the ammunition trolley have round holes. [Marcel Verhaaf Collection]

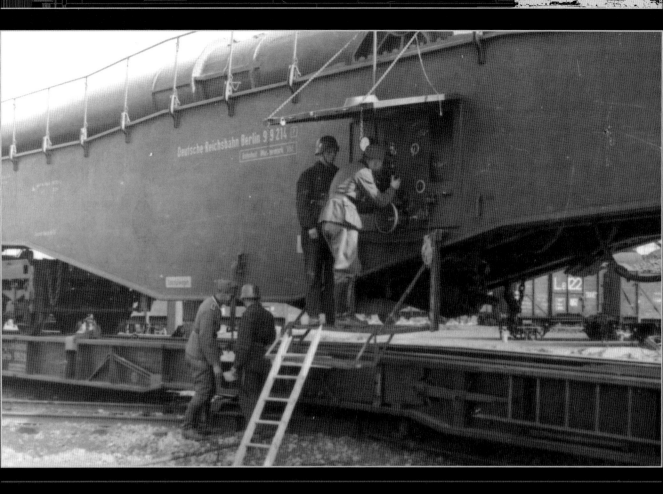

[Continued from page 43]
the pattern seen on both surviving guns. The first pattern lacks the prominent gear casing on the left but instead shows a hand wheel on the right side of the traversing mechanism, used to lock it. Krupp drawings dating from 1937 and 1942 show Ausführung C and D (Models C and D) respectively. They lack the Sprengwerk and both have the later traversing mechanism. Ausführung C has the aiming stand in a low position just as 'Leopold' has. Ausführung D has the aiming stand high, as on the gun in France. This was known as Richtstand Neuer Art, or new model aiming stand. Making an educated guess, I would say the first vehicle built could be Ausführung A, with Sprengwerk and first pattern traversing mechanism. The second and third vehicle with Sprengwerk and second pattern traversing mechanism could be Ausführung B. The fourth gun is consequently Ausführung C. Most of the information on the development of K5 comes from interviews with Krupp employees done by the American Forces at the end of the [Continued on page 47]

[Right] This K5 seems to be in patchy winter camouflage. The function of the fixtures on the barrel is unknown. This photograph may have been taken in the Soviet Union. [Marcel Verhaaf Collection]

[Right] K5 fires a round, the men on the gun are bracing themselves for the recoil. All personnel seem to be wearing overalls. The pivot of the firing bed can be seen resting on the existing rail across which the turntable is built. No carriage number can be read. [Marcel Verhaaf Collection]

[Below] Carriage number 919400 being shown to a group of officers and officials. The ammunition supply car and the ammunition boxcars are in the background. Note the driver's cab on the ammunition supply car. The projectiles and charges were in separate boxcars that had doors in the front and rear, so the ammunition trolley could move through them. One of the large front doors of the first boxcar is open. This K5 has the later aiming stand shown on a drawing dated 1942. Since it seems to be overall dark yellow, it must be 1943 or later. The construction of the aiming stand can be clearly seen. The members of the crew again seem to be wearing overalls. [Marcel Verhaaf Collection]

[Right] A K5 in France, seen in one of the bunkers that were built to hide and protect them. It is in firing mode, since the generator is on the gun. The bunkers were made to look like barns. Note the big sliding doors; with all the rivets, they seem to be steel. Judging from the uniforms, this must be later in the war. [Marcel Verhaaf Collection]

[Bottom] A K5 being shunted by a diesel locomotive. This is most likely a WR360C14. Judging by the house in the background, I would say this photograph was taken in France, on the Channel coast. Behind the train may be one of the bunkers used for concealment. Note the load of unidentified containers on the front railway carriage. The gun is overall dark gray. [Marcel Verhaaf Collection]

[Continued from page 45] war. There is not a lot of documented information left. For this reason, nothing definitive can be said on the number of K5's produced. In a German Army document dated 24 September 1944, there is mention of 10 guns lost, among them 'Leopold' and 'Robert' captured in Italy, and 12 remaining. This leads to a total of 22 guns. Then on 28 September 1944, another document states a total of 36 guns ordered, of which 22 were delivered. The latest documented delivery took place in June 1944, so the total of 22 guns delivered is probably correct.

Deployment

From 1940 onwards, K5's were delivered to the railway batteries. After training they were first deployed on the Channel coast in support of the invasion of Great Britain, called Operation Seelöwe (Sea Lion), which was later cancelled. By the beginning of 1941 there were four Eisenbahnbatterien (railway batteries), 710, 712, 713 and 765, stationed around Boulogne and Calais. They were used to shell [Continued on page 51]

[Left] This is one of a number of photographs taken after the capture of Eisenbahnbatterie 712's guns in Italy. These photos were taken in Civitavecchia where the guns were captured. This particular gun is 'Leopold', carriage number 919219. The cover of the aiming stand is held with struts, whereas that of 'Robert' is held with chains (see photo below). This is a good example of the, undoubtedly many, minor differences between the carriages. The frame for camouflage with tarpaulins is still largely in place, which would give the complete gun the appearance of a line of boxcars. The frame over the barrel has downward extensions to make sides. These were not connected to the railway truck. This setup proves the guns were in transport mode when captured. Removing the tarpaulins was the first task on hand when converting to firing mode. [Daniele Guglielmi Collection]

[Below left] This is 'Robert', carriage number 919215, after its capture. The rammer stand to the side of the breech is missing, just like on 'Leopold'. This is probably consistent with the method used for rendering the guns incapable of firing. Here as well, the frame is still in place. Other photographs show it will disappear over time. Note that the frame attaches to the same brackets as the uprights of the railing. The paint job is very uneven and squiggly. [Photograph NA 16346 courtesy of the Imperial War Museum, London]

[Above right] This is 'Leopold' again. This photograph provides a good overall view of the frame. The standing struts are all angled inward to form the shape of a roof. In the background is a WR360C14 diesel locomotive. It can be recognized from its large radiator. It was part of the normal equipment making up the train for K5. It was used when the gun had to be shunted into its firing position. [Photograph NA 16345 courtesy of the Imperial War Museum, London]

[Right] This photograph shows 'Leopold'. The frames for camouflage with tarpaulins are still in place on the front of the gondola and near the breech. Note the way the railings fold down. The paint seems very uneven and mottled. On the gondola, the white lettering can hardly be distinguished, and dark gray patches have been left around the lettering on the railway truck. The extra structure on the traversing mechanism can be seen behind the support wheel. [Marcel Verhaaf Collection]

[Right] This photograph and the following ones show another captured K5. When and where these photographs were taken is unknown. The soldiers photographed with the gun are British Commonwealth troops. The top attachments for the Sprengwerk are on the cradle, making this an early carriage. A large part of the wooden work surface is missing from the loading platform. There is a single set of rails for the loading trolley. [Jeffrey Plowman Collection via Daniele Guglielmi]

[Far right] Judging from the similarities in the overhead electric lines this is the same gun from the front. It seems in pretty good shape. The tunnel in the background was probably used to hide the gun. [Jeffrey Plowman Collection via Daniele Guglielmi]

[Above] This photograph gives a very good impression of the size of K5's breech. Note the detail on the breech-block. The head of one of the loading ramrods can be seen in the bottom of the photograph. Since the stand for the rammer next to the breech is missing on both guns, it is impossible to say which K5 this is. [Photograph NA 16347 courtesy of the Imperial War Museum, London]

[Right] Another view of 'Leopold' after its capture. The carriage number 919219 can just be read. On the gondola, the new coat was painted around the lettering, which was a common practice when repainting. [Marcel Verhaaf Collection]

[Continued from page 47]
the English Channel coast. Next, they were assigned to different Army Groups for the invasion of Russia. K5's were used in the sieges of Sevastopol, Leningrad and Stalingrad. But they would return to the Channel coast again and again to find their most intensive use there. By 1944, they were part of the Atlantic Wall, with prepared positions to fire from and bunkers for concealment. Eisenbahnbatterie 712 was to be sent to North Africa, but the Germans were beaten before they reached it. They saw action in Italy during the landings at Anzio. Their guns became collectively known as 'Anzio Annie' to the Allied troops. Both guns were captured and 'Leopold' was sent to the USA. Eisenbahnbatterie 712 was reequipped in northern Italy and disbanded in Germany at the end of 1944. The last remaining batteries on the Channel Coast, 688, 710 and 713, were withdrawn to The Netherlands in September 1944 and subse-

quently destroyed there. Battery 686 was withdrawn from the Eastern Front in the middle of 1944 because of a lack of targets and ammunition. This lack of ammunition had plagued K5 throughout its career. There are several reports of guns standing idle for long periods because of it. In some cases, the projected monthly output of 50 projectiles could not be met. Priority was frequently given to the production of commoner types of ammunition. Sometimes the entire output was fired by the guns on the Channel Coast, leaving the other guns idle. When Eisenbahnbatterie 712 fired 523 rounds at the Anzio bridgehead over a period of three months, this led to severe shortages of ammunition on other fronts. Documents mention an unforeseen use of ammunition. On 14 April 1944 there is a total of 1486 grenades available for 16 operational guns.
It makes you wonder if K5 was really an effective weapon. It was best suited to siege warfare, whereas the Second World War had a very mobile

character. It was vulnerable to reconnaissance and attack from the air. For example, Eisenbahnbatterie 712 had to abandon its guns because bombing destroyed the railway. Bombers could do K5's job cheaper and more effectively even in 1940, so we could conclude that K5, for all its technical achievement, was actually obsolete by the time it entered service.

Transport
Railway batteries equipped with K5 had two guns each. Operating these guns required large numbers of people and a lot of equipment. Gun crews consisted of 42 personnel, 25 of whom were employed in operating the gun. Eight of the gunners were stationed on the loading platform and were involved in hoisting up the rounds and loading the gun. The rest were on stations around the gun. A further eight were employed in setting fuzes and transporting the projectiles and propellant charges to the gun. Three

specialists are on hand to operate the generator and maintain the electrical system. Non-commissioned officers oversee aiming, the generator and the ammunition. And of course there are officers in charge of each gun. Beside the gun crews a battery needed a staff, administrative personnel, drivers and mechanics to operate and maintain the trains and motor vehicles, cooks, etc. It took three complete trains to transport a railway battery: one for the staff, one each for the guns. Normal transport used steam locomotives, probably Baureihe (production model) 50 or 52. The latter was a war version, designed to be built economically, both in materials and production time. For deployment, the railway battery used diesel locomotives, since the smoke from steam locomotives tended to give away their position. These steam locomotives may have only been used to pull the trains to their destination. Since they were Reichsbahn equipment, they would

[Right] This K5 has some damage, but it seems largely intact. On most K5s there is a circular hatch on the left front of the gondola, next to the set of handles. An auxiliary motor to operate the gun in case of power loss could be attached here. This vehicle has no such hatch. Only the very first two guns produced didn't have this provision. The guard in the foreground is intriguing as well. He seems to be wearing British battle dress, but with German ammunition pouches and rifle. Maybe this is a member of the resistance. The men on the gun are British Commonwealth troops. [Photograph NA 25193 courtesy of the Imperial War Museum, London]

have been re-assigned. Even the cars making up the train were partly Reichsbahn property.

The staff train had the following composition: steam locomotive plus tender, passenger cars, special box cars with ammunition, box cars with food and field kitchen, a box car with railway equipment, a diesel locomotive, flatbed cars with vehicles, equipment and anti-aircraft guns. The train consisted of 34 cars, including locomotives, with a total length of 374 m (1,215 ft). The gun trains differed slightly from one another. They consist of 24 and 23 cars with a length of 294 and 278 m (956 and 904 ft) respectively. They had both steam and diesel locomotives, a K5, a Munitions-zubringerwagen (ammunition supply car), which was used to transport the generator unit, a set of three box cars for the ammunition, a flatbed car with field kitchen, anti-aircraft guns, a special car for the firing bed (Schiessbettung) and a car with a crane and turntable parts. The longer train had four, instead of three, passenger cars and a box car with food. The shorter one had two box cars for the armourer (Waffen-meister). In combat configuration, most of the passenger cars and cargo cars would be uncoupled from the train. On a drawing of 24 August 1937 combat configuration was as follows: diesel locomotive, two flatbed cars, K5, ammunition supply car with generator, ammunition car 1, ammunition car 2, a box car with temperature control and auxiliary generator, equipment car and a workshop. These are of course official compositions. Whether or not they were followed to the letter in practice is unknown. This smaller train was shunted into its firing position with the diesel locomotive. The turntable was prepared beforehand.

Preparing for firing

In transport, K5 and the Vögele Drehscheibe (turntable) were concealed under tarpaulins over a framework, to make them look like normal boxcars. On arrival, the tarpaulins and frames would be taken off first and laid to the side. The

[Right] In this and the following photographs, a distinct difference in color can be seen between the carriage itself and the barrel. Note that the cradle has the same color as the carriage. Here is visible proof of the barrel being replaced when it was worn out or the particular ammunition it fired ran out. This barrel change must have taken place after the 1943 replacement of dark gray with dark yellow. No effort was spent on repainting the carriage itself. [Jeffrey Plowman Collection via Daniele Guglielmi]

handbrake of the front railway truck was applied. The chain connecting the rear railway truck and the gondola of the gun was unhooked and hung on the gondola. A three-way valve next to the aiming stand regulated the compressed air system. It was moved from 'transport mode' to 'closed', and then the pressure hose was closed, and uncoupled from an angle cock beneath the aiming stand. The front locking system of the gondola was unlocked, and it was raised slightly by hydraulics to make moving it

easier. From the transport position, the rear railway truck was moved backwards 1.9 m (6 ft) to its firing position and locked in place with the rear locking mechanism. The rear railway truck could be moved either by locomotive or by hand. The pressure hose was then reconnected to a valve further back on the gondola. The three-way valve was set to 'firing mode' (Ausgefahrenes Drehgestell or railway truck in rear position). Firing mode didn't make the gun immobile. The extra length would make it harder to operate on

the railway for long-distance transport. The generator had to be moved from its transport position on the ammunition supply car to the rear railway truck. Rails were fixed between them. The generator was jacked up and put on rollers, one side at the time. It was then pulled onto the rear railway truck by means of pulleys attached to eyes on the back of the gondola. The rollers were removed and the generator secured to the rear railway truck. The electrical connections were then made. Meanwhile, the railings would

have been erected, ladders extended or attached, the barrel clamp released, the cover removed from the barrel, the aiming stand put into operation and all equipment needed to operate the gun put in place. K5 could either be fired from a curved length of railway (Schiesscurve) or from the Vögele Drehscheibe. In the first case, the gun would be aimed roughly by shunting forward or backward. Hand brakes were applied and a ruler was attached to the track forward of the aiming stand to measure backslide caused by

[This page, all photos] More photos of the same gun. Since Eisenbahn-batterie 749 was overrun in the south of France near Montelimar, this may be one of their guns, though not the one preserved in Audinghen, which is a late production model. [Top: photograph NA 25194 courtesy of the Imperial War Museum, London; top right and right: Jeffrey Plowman Collection via Daniele Guglielmi]

[Right and bottom] These photos were taken in a railway station. The difference in color between the carriage and the barrel is obvious here. [Jeffrey Plowman Collection via Daniele Guglielmi]

recoil. The gun could then be accurately repositioned after firing. Ammunition cars would be placed away from the gun, fuzes were set here and charges prepared. Projectiles and charges were stowed in separate cars. They were loaded onto ammunition trolleys, with the projectile in the middle and the charges stowed to the sides. Rails ran along the length of the ammunition cars and between them, the latter movable, to allow the trolley to move through them. The ammunition supply car was used to bring the trolleys to the gun. It ope- rated under its own power. The Vögele Company produced turn- tables for conventional railway purposes; these were used, for in- stance, to turn locomotives around in railway yards. The Vögele Drehscheibe was conceived bet- ween the two World Wars, during the same studies that produced the theoretical basis for the develop- ment of K5. It consists of a circular rail and a firing bed, to give a 360- degree arc of fire. It was used with other railway guns as well. The firing bed rotated on a pivot attached to the railway and the circular rail. A com- plete K5 was moved onto the firing bed and secured to recoil and recuperator cylinders on the front of the firing bed. The large coupling is used for this. The gun's generator also powered the firing bed. When using the turntable, ammunition trolleys were transferred from the ammunition supply car to a special car running on the rail of the turn- table to get them to the gun. The trolley was hoisted to the loading platform and rolled to the breech. The grenade was loaded using a rammer. The charge consisted of several bags of propellant, the main one in a metal cartridge. The bags and the cartridge were put in the chamber with another rammer. Both rammers were stored on the loading platform. The breech could now be closed. The loading flap was raised and the gun was aimed from the aiming stand and fired by means of a lanyard. Officially all personnel had to be off the gun when firing. After firing the cartridge was caught on the cartridge catcher that swiveled to the side in order to reload the gun. The gun was reloaded and the empty cartridge was put on the empty trolley and hoisted to the ground.

[Right] Cleaning the barrel of a K5. It has been slightly elevated to make the work easier. The cover of the aiming stand, from which the elevation motors were operated, is raised. Whether the gun is in firing mode is unclear. The bit of roof on the far left appears to belong to a diesel locomotive (see photo below). Since they were only used when the guns were near or in their firing positions this would point to firing mode. In the background is another tunnel, which is entirely consistent with the fact that tunnels were often used to conceal railway guns. Note the lettering on the flatbed car. The shelter seems to be wood; the men can use the roof as a platform to perform their task. [Daniele Guglielmi Collection]

Camouflage

The first four guns were overall dark gray when they left the factory in 1937. They probably received camouflage when they were tested. Most of the period photographs I have seen showing this camouflage also show provisions for the Sprengwerk that was used only on the first three guns. The two-tone scheme was in use from 1935 to 1939 and consisted of 2/3 dark gray and 1/3 brown. On K5, it was applied in broad undulating vertical bands (see photograph on page 41). By the time K5's were delivered to the troops, the color scheme had changed to overall dark gray. They were sent to France and Russia in this livery. During the Russian winter, K5's may have been whitewashed to camouflage them (see photograph on page 45, bottom). The Afrikakorps used a dark desert yellow from 1942 onwards. Since 'Leopold' was destined for North Africa it may have had this color. In early 1943, a three-tone camouflage scheme was adopted, using a base of Dunkelgelb (dark yellow). Dark brown and green could be added according to season and terrain. K5 guns seem to be found only in overall dark yellow, although the painting could be very motley. Lettering and badges were often left in patches of the original dark gray, as is the case on 'Leopold'.

Charges and projectiles

Ammunition for K5 was of the separate loading type. The complete round consisted of: the projectile with fuze, three separate charges in cloth bags and the main charge in a brass cartridge case with a primer in the base. This was the maximum charge; fewer bags would be used for shorter ranges. All of these charges were made up of a core that ignited the next charge, surrounded by rods of propellant. Apart from the main charge, they were capped and then contained in white cloth bags. These charges are called Vorkartusche (pre-charge). The large ones, 1. and 2. Vorkartusche, contained 50 kg (110 lbs) of propellant. The small one, 3. Vorkartusche, contained just 21 kg (46 lbs). In the drawings the cloth bags look shaped to show the contours of the charge inside, in reality they were straight

[Left] A WR360C14 diesel locomotive seen from the side. This was the most numerous German diesel locomotive. They were used for shunting purposes on all fronts, both by the Wehrmacht and the Reichsbahn. For work in fuel and ammunition dumps, diesel locomotives were preferred to steam locomotives that produced sparks, as they were safer. They were also used to shunt K5's to their firing positions since they do not produce huge clouds of smoke and steam. They would thus be less conspicuous from the air. [Marcel Verhaaf Collection]

[Right] The ammunition crew showing their stuff. The ammunition trolley carries the maximum charge, consisting of two large and one small bags of propellant, plus the main charge in the cartridge case. Altogether they weighed about 250 kg (552 lb). The projectile adds another 255.5 kg (564 lb) to the load. The men are on the ammunition supply car. Note that there are three rails; the generator unit ran on the outer ones. [Marcel Verhaaf Collection]

tubes with stiff top and bottom. The main charge of 60 kg (132 lb), in its cloth bag, was loaded into the cartridge and then capped to secure it. This is the Hauptkartusche. Stenciled lettering appeared only on the cloth bags, including that of the main charge. The lettering runs across the charge in the drawings, but in period photographs I have seen, it runs lengthwise. Both practices may have been in use. The cartridge case may have had stenciled codes on the base, but none can be seen in period photographs. They would have been reused, only the percussion primer had to be replaced. Maybe only this item carried codes. Codes for checking were on the cloth bag of the main charge.

K5 fired several types of ammuni-tion, partly dictated by the different types of barrel, but also because special types were used for practice and calibration of the gun. Barrels with deep rifling (Tiefzug T10 or T7) used projectiles with rotating ribs. The height of these ribs was either 10 or 7 mm; they were made of cast iron. The standard high-explosive projectile was the 28 cm Granate 35 (28 cm grenade 35). There was also a rocket assisted round called the 28 cm R-Granate 4331. The R stands for Raketenantrieb (rocket pro-pelled). The top of the projectile held an additional propellant charge that was ignited after 19 seconds to give greater range. These rounds, both projectile and charge, had to be kept within a temperature range of +5 to +15 degrees Celsius. Which also explains the necessity of tempe-rature control equipment with the ammunition carriages.

Barrels with multiple rifling (Vielzug) used conventional projectiles with rotating bands. The standard projectile was the 28 cm Granate 42 (28 cm grenade 42) and its rocket-assisted counterpart 28 cm R-Granate 4341. Performance and re-strictions are as above. Because of continual shortages of ammunition, 28 cm projectiles for Bruno railway guns were reworked for K5 and designated 28 cm Granate 39/42.

The last projectile to be developed for K5 was the 31 cm Spreng-granate 4861 (31 cm high-explosive grenade 4861) also known as Peenemünder Pfeil Geschoss (arrow projectile). This is a sub-caliber projectile with a guide ring in the middle and stabilizing fins on the tail. it was fired from a smoth bore barrel. Its development was plagued by setbacks and concluded very late in the war. Whether any were fired in anger is uncertain (see page 43). Per original drawings, the projectiles were painted Feldgrau (field gray), with white lettering for type of fuze, type of projectile and weight of projectile. In period photographs some of the lettering is shown black and some white. Lettering between the ribs is white; the rest seems to be black, apart from large numbers 15 or 16 on the cone of the projectile. The cones appear very shiny some-times, so they may have been painted gloss. The copper rotating bands on the projectile would remain unpainted. ☐